Copyright © 2011 XAMonline, Inc.
All rights reserved. No part of the material protected by this copyright notice may be reproduced or utilized in any form or by any means, electronic or mechanical, including photocopying, recording or by any information storage and retrievable system, without written permission from the copyright holder.

To obtain permission(s) to use the material from this work for any purpose including workshops or seminars, please submit a written request to:

XAMonline, Inc.
25 First Street, Suite 106
Cambridge, MA 02141
Toll Free: 1-800-509-4128
Email: info@xamonline.com
Web: www.xamonline.com
Fax: 1-617-583-5552

Library of Congress Cataloging-in-Publication Data

Wynne, Sharon A.
    MTEL Mathematics 09 Practice Test 2: Teacher Certification /
Sharon A. Wynne. -1st ed.
    ISBN: 978-1-60787-212-2
    1. MTEL Mathematics 09 Practice Test 2
    2. Study Guides   3. MTEL   4. Teachers' Certification & Licensure
    5. Careers

Disclaimer:
The opinions expressed in this publication are the sole works of XAMonline and were created independently from the National Education Association, Educational Testing Service, or any State Department of Education, National Evaluation Systems or other testing affiliates.

Between the time of publication and printing, state specific standards as well as testing formats and website information may change that is not included in part or in whole within this product. Sample test questions are developed by XAMonline and reflect similar content as on real tests; however, they are not former tests. XAMonline assembles content that aligns with state standards but makes no claims nor guarantees teacher candidates a passing score. Numerical scores are determined by testing companies such as NES or ETS and then are compared with individual state standards. A passing score varies from state to state.

Printed in the United States of America   œ-1
MTEL Mathematics 09 Practice Test 2
ISBN: 978-1-60787-212-2

Mathematics
Post-Test Sample Questions

1. What is the smallest number that is divisible by 3 and 5 and leaves a remainder of 3 when divided by 7?
   (Average)

   A. 15

   B. 18

   C. 25

   D. 45

2. Which of the following is an equivalent representation of $\dfrac{3-4i}{1+2i}$?
   (Average)

   A. 3

   B. $2 - 6i$

   C. $3 - 2i$

   D. $-1 - 2i$

3. What is the GCF of 143 and 156?
   (Average)

   A. 2

   B. 3

   C. 13

   D. No common factors

4. What would be the total cost of a suit for $295.99 and a pair of shoes for $69.95 including 6.5% sales tax?
   (Average)

   A. $389.73

   B. $398.37

   C. $237.86

   D. $315.23

5. Which graph shows the solution to the system of inequalities below?

$3x - 2y \leq 5$
$-x + 5y > 1$

(Rigorous)

A)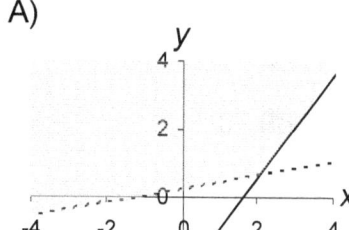

B)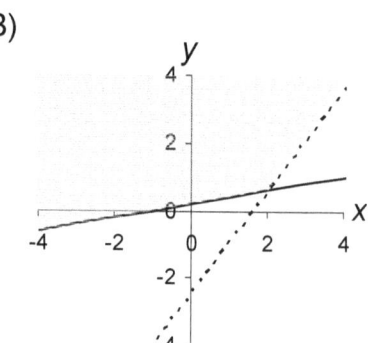

C)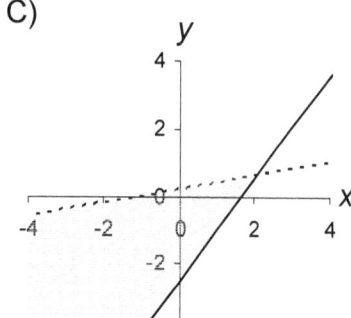

D)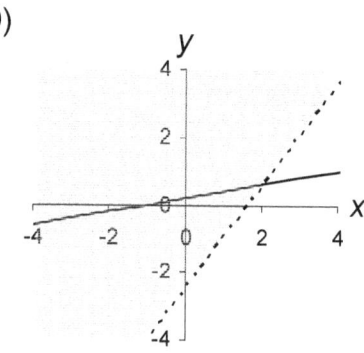

6. What are the foci of the ellipse $16(y-3)^2 = 16 - (x-2)^2$?
   (Rigorous)

   A. $(2 \pm 4, 3)$

   B. $(-2 \pm \sqrt{15}, -3)$

   C. $(2 \pm \sqrt{15}, 3)$

   D. $(2, 3 \pm \sqrt{15})$

7. Which unit of measurement would be the most appropriate for characterizing the weight of a dime?
   (Easy)

   A. Gram

   B. Kilogram

   C. Pound

   D. Ton

8. A scientist is measuring a physical constant that has an accepted value of 5.729 units. If the scientist's measurement is 5.693, what is his percent error?
   (Average)

   A. 0.0063%

   B. 0.63%

   C. 1.79%

   D. 10%

9. An archer's paper target shows the hits illustrated below. Which term best describes the archer's shooting in this case?

   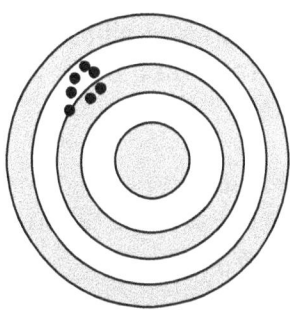

   (Average)

   A. Accurate

   B. Precise

   C. Exact

   D. On target

10. Which theorem can be used to prove $\triangle BAK \cong \triangle MKA$?
    (Average)

    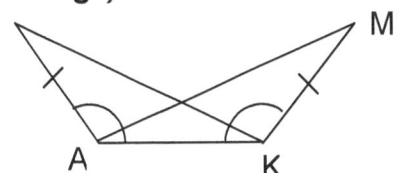

    A. SSS

    B. ASA

    C. SAS

    D. AAS

11. Choose the correct statement concerning the median and altitude in a triangle. *(Average)*

    A. The median and altitude of a triangle may be the same segment

    B. The median and altitude of a triangle are always different segments

    C. The median and altitude of a right triangle are always the same segment

    D. The median and altitude of an isosceles triangle are always the same segment

12. What is the measure of minor arc AD, given measure of arc PS is 40° and $m\angle K = 10°$? *(Rigorous)*

    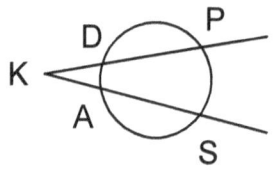

    A. 50°

    B. 20°

    C. 30°

    D. 25°

13. Determine the area of the shaded region of the trapezoid in terms of *x* and *y*. *(Rigorous)*

    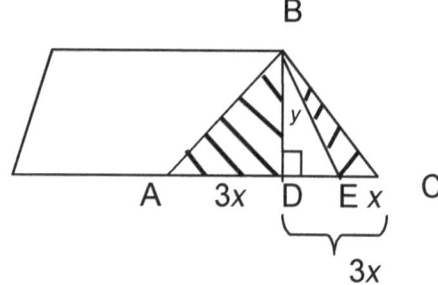

    A. $4xy$

    B. $2xy$

    C. $3x^2 y$

    D. There is not enough information given

14. Two points have coordinates (3, –4, 1) and (6, 2, –7). What is the distance between these points? *(Easy)*

    A. 7 units

    B. 10.4 units

    C. 13.5 units

    D. 15 units

15. Given $K(-4, y)$ and $M(2, -3)$ with midpoint $L(x, 1)$, determine the values of x and y.
    (Rigorous)

    A. $x = -1, y = 5$

    B. $x = 3, y = 2$

    C. $x = 5, y = -1$

    D. $x = -1, y = -1$

16. The cosine function is equivalent to
    (Easy)

    A. $\dfrac{1}{\text{sine}}$

    B. $\dfrac{1}{\text{tangent}}$

    C. $\dfrac{\text{sine}}{\text{tangent}}$

    D. $\dfrac{\text{cotangent}}{\text{sine}}$

17. Determine the measure of the angle α in the triangle below.
    (Rigorous)

    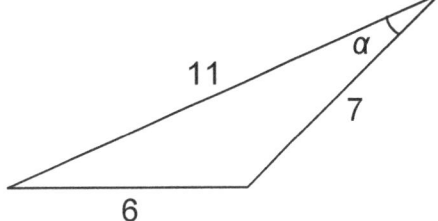

    A. 15.3°

    B. 18°

    C. 29.5°

    D. 45°

18. For an acute angle x, sin x = 0.6. What is cot x?
    (Rigorous)

    A. $\dfrac{5}{3}$

    B. 0.75

    C. 1.33

    D. 1

19. Determine the rectangular coordinates of the point with polar coordinates (5, 60°).
    (Average)

    A. (0.5, 0.87)

    B. (−0.5, 0.87)

    C. (2.5, 4.33)

    D. (25, 150°)

20. A population P of bacteria doubles in number every hour. Which of the following functions of t in hours best represents the number of bacteria in the population?
    (Rigorous)

    A. P^t

    B. $P(2^t)$

    C. Pt^2

    D. Pe^t

21. What is the domain of the function $g(x) = \tan x$?
    (Average)

    A. $\{x \in \Box\}$

    B. $\{x \in \Box : x \neq \pm n\pi\}$ (n odd)

    C. $\left\{x \in \Box : x \neq \pm \dfrac{n\pi}{2}\right\}$ (n odd)

    D. $\{\varnothing\}$

22. Find the zeroes of $f(x) = x^3 + x^2 - 14x - 24$
    (Rigorous)

    A. 4, 3, 2

    B. 3, −8

    C. 7, −2, −1

    D. 4, −3, −2

23. Which equation corresponds to the logarithmic statement: $\log_x k = m$?
    (Rigorous)

    A. $x^m = k$

    B. $k^m = x$

    C. $x^k = m$

    D. $m^x = k$

24. Which expression is equal to $x^4 + 2x^3 - 16x^2 - 2x + 15$ divided by $x + 5$?
    (Rigorous)

    A. $4x^3 + 6x^2 - 32x - 2$

    B. $x^3 - 15x^2 - 5x + 15$

    C. $x^3 - 3x^2 - x + 3$

    D. $5x^3 + 10x^2 - 80x + 75$

25. Solve for x: $10^{x-3} + 5 = 105$.
    (Rigorous)

    A. 3

    B. 10

    C. 2

    D. 5

26. Find the inverse of the function $f(x) = 2x^2 - 3$.
    (Average)

    A. $f^{-1}(x) = \sqrt{\dfrac{x+3}{2}}$

    B. $f^{-1}(x) = 2x^2 + 3$

    C. $f^{-1}(x) = \sqrt{2x^2 + 3}$

    D. The function does not have an inverse

27. Which of the following represents $f \circ g$, where $f(x) = 3x^2 + 1$ and $g(x) = 2\sin x - 1$?
    (Average)

    A. $2\sin(3x^2 + 1) - 1$

    B. $6\sin x - 2$

    C. $3\sin x + 1$

    D. $12\sin^2 x - 12\sin x + 4$

28. Which of the following functions does not have an inverse?
    (Average)

    A. x^3

    B. $\ln \dfrac{x}{2}$

    C. e^{x^2}

    D. $\dfrac{1}{x}$

29. Find the following limit:
    $$\lim_{x \to 2} \frac{x^2 - 4}{x - 2}$$
    (Average)

    A. 0

    B. Infinity

    C. 2

    D. 4

30. Find the following limit:
    $$\lim_{x \to 0} \frac{\sin 2x}{5x}$$
    (Rigorous)

    A. Infinity

    B. 0

    C. 1.4

    D. 1

31. The radius of a spherical balloon is increasing at a rate of 2 feet per minute. What is the rate of increase of the volume when the radius is 4 feet?
   (Rigorous)

   A. 4 feet³/minute

   B. 32π feet³/minute

   C. 85.3π feet³/minute

   D. 128π feet³/minute

32. What is the maximum value of the function $f(x) = -3x^2 - 5$?
   (Average)

   A. −5

   B. −3

   C. 1

   D. 5

33. Find the antiderivative for the function $y = e^{3x}$.
   (Rigorous)

   A. $3x(e^{3x}) + C$

   B. $3(e^{3x}) + C$

   C. $1/3(e^x) + C$

   D. $1/3(e^{3x}) + C$

34. Evaluate $\int_0^2 (x^2 + x - 1)dx$
   (Rigorous)

   A. 11/3

   B. 8/3

   C. −8/3

   D. −11/3

35. What conclusion can be drawn from the graph below?

   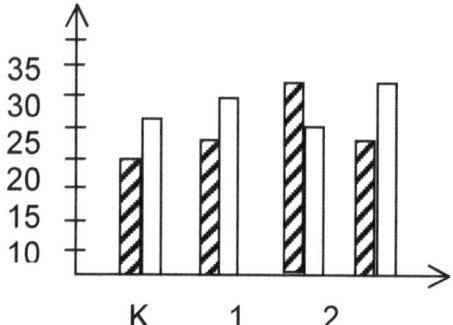

   MLK Elementary Student Enrollment
   Girls Boys
   (Easy)

   A. The number of students in first grade exceeds the number in second grade

   B. There are more boys than girls in the entire school

   C. There are more girls than boys in the first grade

   D. Third grade has the largest number of students

36. The pie chart below shows sales at an automobile dealership for the first four months of a year. What percentage of the vehicles were sold in April?
    (Easy)

    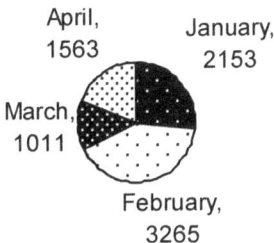

    A. More than 50%

    B. Less than 25%

    C. Between 25% and 50%

    D. None

37. What is the standard deviation of the following sample?

    {1.46, 1.55, 1.55, 1.57, 1.89, 2.01, 2.09}

    (Average)

    A. 0.010

    B. 0.066

    C. 0.26

    D. 1.73

38. Which measure of central tendency best characterizes the data set below?

Value	Frequency
1	2
2	5
3	7
4	3
5	2
6	1
7	1

    (Rigorous)

    A. Mean

    B. Median

    C. Both the mean and median are the same

    D. None of the above

39. The height of people in a certain city is a normally distributed random variable. If a person is chosen from the city at random, what is the probability that he or she has a height greater than the mean of the distribution?
    (Average)

    A. 0.25

    B. 0.5

    C. 0.75

    D. Not enough information

40. If there are three people in a room, what is the probability that at least two of them will share a birthday? (Assume a year has 365 days.)
    (Rigorous)

    A. 0.67

    B. 0.05

    C. 0.008

    D. 0.33

41. How many different five-card hands containing three aces and two kings can be drawn from a standard 52-card deck?
    (Rigorous)

    A. 6

    B. 16

    C. 24

    D. 2,598,960

42. What is the probability that a roll of a six-sided die yields an outcome that is even or greater than three?
    (Average)

    A. 3/6

    B. 4/6

    C. 5/6

    D. 1

43. Find the sum of the following matrices is

    $\begin{pmatrix} 6 & 3 \\ 9 & 15 \end{pmatrix} \begin{pmatrix} 4 & 7 \\ 1 & 0 \end{pmatrix}$

    (Easy)

    A. $\begin{pmatrix} 10 & 10 \\ 10 & 15 \end{pmatrix}$

    B. $\begin{pmatrix} 13 & 7 \\ 9 & 16 \end{pmatrix}$

    C. 45

    D. $\begin{pmatrix} 20 \\ 25 \end{pmatrix}$

44. The product of two matrices can be found only if
    (Easy)

    A. The number of rows in the first matrix is equal to the number of rows in the second matrix

    B. The number of columns in the first matrix is equal to the number of columns in the second matrix

    C. The number of columns in the first matrix is equal to the number of rows in the second matrix

    D. The number of rows in the first matrix is equal to the number of columns in the second matrix

45. Which of the following properties does not apply to matrix multiplication?
    (Easy)

    A. Associativity

    B. Commutativity

    C. Distributivity

    D. All of the above

46. Which of the following best describes the translation matrix below for arbitrary points (x_1, y_1)?

    $$\begin{pmatrix} 1 & 0 \\ 0 & -1 \end{pmatrix} \begin{pmatrix} x_1 \\ y_1 \end{pmatrix} = \begin{pmatrix} x_2 \\ y_2 \end{pmatrix}$$

    (Average)

    A. Rotation

    B. Translation

    C. Reflection

    D. Dilation

47. The Fibonacci numbers are defined recursively using the expression $F_{i+1} = F_i + F_{i-1}$. If $F_1 = 0$ and $F_2 = 1$, what is F_8?
    (Easy)

    A. 0

    B. 1

    C. 13

    D. 21

48. On which of the following sets is multiplication *not* symmetric?
    (Average)

    A. Real numbers

    B. Complex numbers

    C. Polynomials

    D. Matrices

49. Which of the following can be used to prove that $0.\overline{9} = 1$?
    (Rigorous)

    A. Arithmetic series

    B. Geometric series

    C. Fibonacci sequence

    D. Irrational numbers

50. What is the shortest path between points A and B?

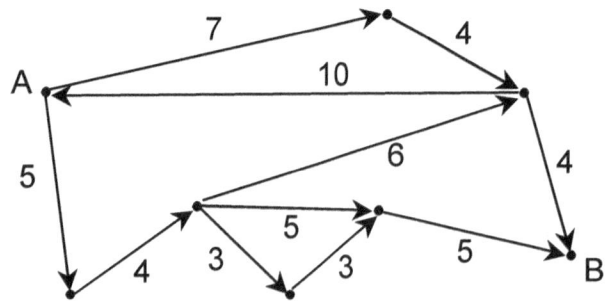

(Easy)

A. 11

B. 14

C. 15

D. 19

Mathematics
Post-Test Sample Questions with Rationales

1. What is the smallest number that is divisible by 3 and 5 and leaves a remainder of 3 when divided by 7?
   (Average)

   A. 15

   B. 18

   C. 25

   D. 45

Answer: D. 45
To be divisible by both 3 and 5, the number must be divisible by 15. Inspecting the first few multiples of 15, you will find that 45 is the first of the sequence that is 4 greater than a multiple of 7.

2. Which of the following is an equivalent representation of $\dfrac{3-4i}{1+2i}$?
   (Average)

   A. 3

   B. $2 - 6i$

   C. $3 - 2i$

   D. $-1 - 2i$

Answer: D. $-1 - 2i$
Multiply both the numerator and denominator by the complex conjugate of the denominator $(1 + 2i)$ to simplify this complex division.

$$\frac{3-4i}{1+2i} \cdot \frac{1-2i}{1-2i} = \frac{3-6i-4i+8i^2}{1-2i+2i-4i^2} = \frac{-5-10i}{1+4} = \frac{-5-10i}{5} = -1-2i$$

Thus, the correct answer is D.

3. **What is the GCF of 143 and 156?**
   (Average)

   A.   2

   B.   3

   C.   13

   D.   No common factors

Answer: C. 13
One way to determine the greatest common factor is to find the prime factorization (the factorization of the number in terms of prime numbers only) of each number. A strategy for prime factorization for small numbers such as those given in this question is to test each prime, starting with 2 and increasing, as a factor of the number.

$156 = 2 \cdot 78 = 2 \cdot 2 \cdot 39 = 2 \cdot 2 \cdot 3 \cdot 13$
$143 = 11 \cdot 13$

Note that these two numbers share only one common factor: 13. Thus, 13 is the GCF. The correct answer is C.

4. **What would be the total cost of a suit for $295.99 and a pair of shoes for $69.95 including 6.5% sales tax?**
   (Average)

   A. $389.73

   B. $398.37

   C. $237.86

   D. $315.23

Answer: A. $389.73
Before the tax, the total comes to $365.94. Then .065($365.94) = $23.79. With the tax added on, the total bill is $365.94 + $23.79 = $389.73. (A quicker way is 1.065($365.94) = $389.73.)

5. Which graph shows the solution to the system of inequalities below?

   $3x - 2y \leq 5$
   $-x + 5y > 1$

   (Rigorous)

A.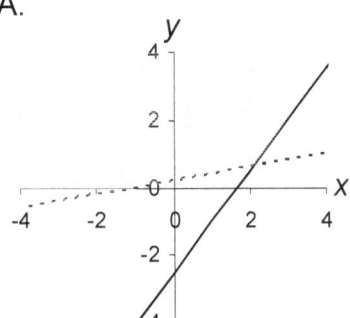

B.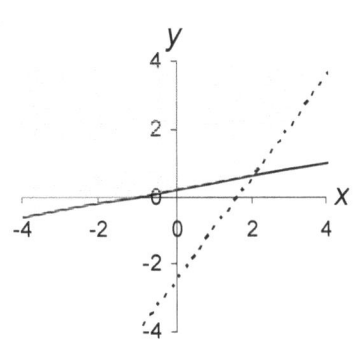

C.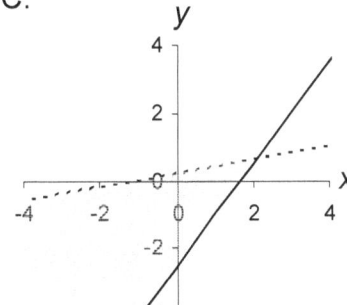

D.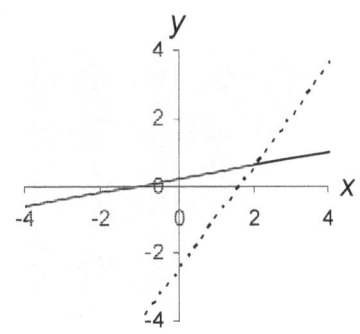

Answer: A.
First, solve each inequality for *y*, as shown below.

$3x - 2y \leq 5$ $-x + 5y > 1$
$2y + 5 \geq 3x$ $5y > x + 1$
$2y \geq 3x - 5$ $y > 0.2x + 0.2$
$y \geq 1.5x - 2.5$

These two linear inequalities can be plotted separately on the same graph. Recall that the boundary line of the solution set is found by replacing the inequality symbol (≥ or >, in this case) with an equality. If the inequality is absolute (> or <), a dashed line is used, since the points on the line do not satisfy the inequality. Otherwise, a solid line is used. The appropriate region can then be shaded.

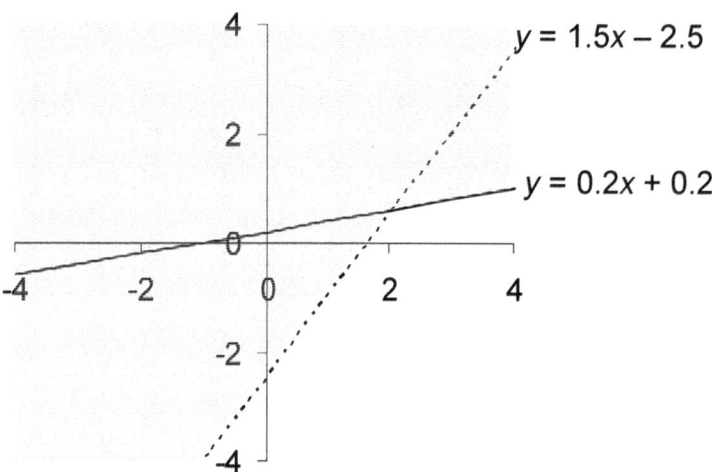

The darkest region is the set of solutions that satisfies both inequalities. Thus, answer A is correct.

6. What are the foci of the ellipse $16(y-3)^2 = 16-(x-2)^2$?
   (Rigorous) (Skill 1.8)

   A. $(2\pm 4, 3)$

   B. $(-2\pm\sqrt{15}, -3)$

   C. $(2\pm\sqrt{15}, 3)$

   D. $(2, 3\pm\sqrt{15})$

Answer: C. $(2\pm\sqrt{15}, 3)$

We can start by attempting to express the equation in standard form for an ellipse:

$$\frac{(x-h)^2}{a^2} + \frac{(y-k)^2}{b^2} = 1$$

Here, (h, k) is the center of the ellipse, and a and b are half the axis lengths. (Depending on the particular values, a or b can be half of the minor or major axes.)

$$16(y-3)^2 = 16-(x-2)^2$$
$$(y-3)^2 = 1 - \frac{(x-2)^2}{16}$$
$$\frac{(x-2)^2}{16} + \frac{(y-3)^2}{1} = 1$$

On the basis of comparison with the standard form, we can quickly see that the center of the ellipse is located at (2, 3). The major axis is parallel to the x-axis, and it has a half-length of 4 units (the square root of 16). We know that the foci are located on the major axis. Their locations are $(h\pm c, k)$, where

$$c = \sqrt{a^2 - b^2}$$

In this case,

$$c = \sqrt{16-1} = \sqrt{15}$$

The foci are then located at $\left(2 \pm \sqrt{15}, 3\right)$.

7. Which unit of measurement would be the most appropriate for characterizing the weight of a dime?
   (Easy) (Skill 2.1)

   A. Gram

   B. Kilogram

   C. Pound

   D. Ton

Answer: A. Gram
A dime is fairly small and obviously doesn't weigh close to a pound. Thus, neither pound, kilogram, nor ton is an appropriate measurement for a dime. The correct answer is A.

8. A scientist is measuring a physical constant that has an accepted value of 5.729 units. If the scientist's measurement is 5.693, what is his percent error?
   (Average)

   A. 0.0063%

   B. 0.63%

   C. 1.79%

   D. 10%

Answer: B. 0.63%
To calculate the percentage error P, divide the absolute difference between the theoretical (accepted) value T and the experimental value E by the accepted value T, then multiply by 100%.

$$P = \frac{|E-T|}{T} \cdot 100\% = \frac{|5.693 - 5.729|}{5.729} \cdot 100\% \approx 0.63\%$$

9. An archer's paper target shows the hits illustrated below. Which term best describes the archer's shooting in this case?

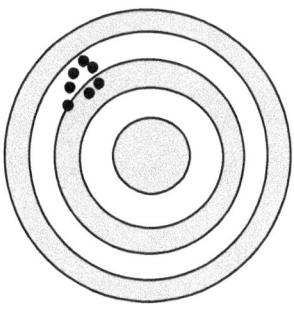

(Average)

   A.   Accurate

   B.   Precise

   C.   Exact

   D.   On target

Answer: B. Precise
The archer shows *precision* in his shooting, in that the hits are tightly grouped in a small area. Because that group is off center, however, the archer is not *accurate*. The terms "exact" and "on target" do not apply in this case. Thus, the correct answer is B.

10. Which theorem can be used to prove $\triangle BAK \cong \triangle MKA$?
    (Average)

    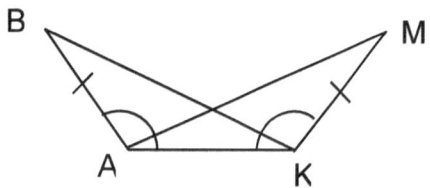

    A. SSS

    B. ASA

    C. SAS

    D. AAS

Answer: C. SAS
Since side AK is common to both triangles, the triangles can be proved congruent by using the Side-Angle-Side Postulate.

11. **Choose the correct statement concerning the median and altitude in a triangle.**
    (Average)

    A. The median and altitude of a triangle may be the same segment

    B. The median and altitude of a triangle are always different segments

    C. The median and altitude of a right triangle are always the same segment

    D. The median and altitude of an isosceles triangle are always the same segment

Answer: A. The median and altitude of a triangle may be the same segment
The most one can say with certainty is that the median (segment drawn to the midpoint of the opposite side) and the altitude (segment drawn perpendicular to the opposite side) of a triangle <u>may</u> coincide, but they more often do not. In an isosceles triangle, the median and the altitude to the <u>base</u> are the same segment.

12. **What is the measure of minor arc AD, given measure of arc PS is 40° and $m\angle K = 10°$?**
    (Rigorous)

    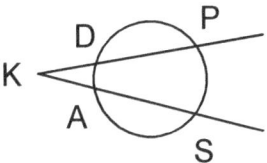

    A. 50°

    B. 20°

    C. 30°

    D. 25°

Answer: B. 20°
The formula relating the measure of angle K and the two arcs it intercepts is $m\angle K = \frac{1}{2}(mPS - mAD)$. Substituting the known values yields $10 = \frac{1}{2}(40 - mAD)$. Solving for *mAD* gives an answer of 20 degrees.

13. Determine the area of the shaded region of the trapezoid in terms of x and y.
    (Rigorous)

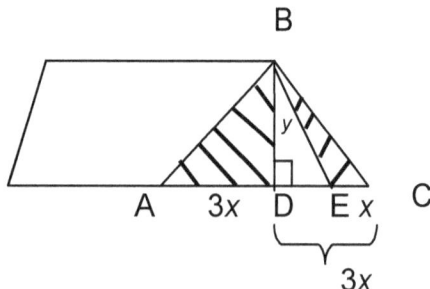

A. $4xy$

B. $2xy$

C. $3x^2y$

D. There is not enough information given.

Answer: B. $2xy$

To find the area of the shaded region, find the area of triangle ABC and then subtract the area of triangle DBE. The area of triangle ABC is $.5(6x)(y) = 3xy$. The area of triangle DBE is $.5(2x)(y) = xy$. The difference is $2xy$.

14. Two points have coordinates (3, –4, 1) and (6, 2, –7). What is the distance between these points?
    (Easy)

    A. 7 units

    B. 10.4 units

    C. 13.5 units

    D. 15 units

Answer: B. 10.4 units
The locations of these points are expressed using three-dimensional coordinates. The distance d can be calculated using the distance formula as follows.

$$d = \sqrt{(x_1-x_2)^2+(y_1-y_2)^2+(z_1-z_2)^2} = \sqrt{(3-6)^2+(-4-2)^2+(1+7)^2}$$
$$d = \sqrt{(-3)^2+(-6)^2+(8)^2} = \sqrt{9+36+64} = \sqrt{109} \approx 10.4$$

15. Given $K(-4,y)$ and $M(2,-3)$ with midpoint $L(x,1)$, determine the values of x and y.
    (Rigorous)

    A. $x=-1, y=5$

    B. $x=3, y=2$

    C. $x=5, y=-1$

    D. $x=-1, y=-1$

Answer: A. $x=-1, y=5$
The formula for finding the midpoint (a,b) of a segment passing through the points (x_1,y_1) and (x_2,y_2) is $(a,b) = (\frac{x_1+x_2}{2}, \frac{y_1+y_2}{2})$. Setting up the corresponding equations from this information yields $x = \frac{-4+2}{2}$ and $1 = \frac{y-3}{2}$. Solving for x and y yields x = –1 and y = 5.

16. The cosine function is equivalent to
    (Easy)

    A. $\dfrac{1}{\text{sine}}$

    B. $\dfrac{1}{\text{tangent}}$

    C. $\dfrac{\text{sine}}{\text{tangent}}$

    D. $\dfrac{\text{cotangent}}{\text{sine}}$

Answer: C. $\dfrac{\text{sine}}{\text{tangent}}$

The cosine function is clearly not the reciprocal of the sine or tangent functions. Simplify answers C or D to determine which is the correct answer. For instance:

$$\dfrac{\text{sine}}{\text{tangent}} = (\text{sine})(\text{cotangent}) = \text{sine}\left(\dfrac{\text{cosine}}{\text{sine}}\right) = \text{cosine}$$

Thus, answer C is correct.

17. Determine the measure of the angle α in the triangle below.

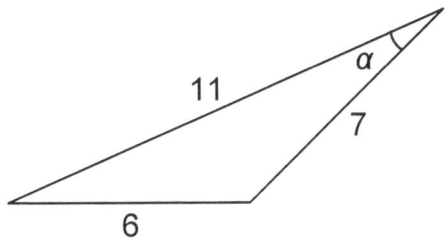

(Rigorous) (Skill 4.2)

A. 15.3°

B. 18°

C. 29.5°

D. 45°

Answer: C. 29.5°
We can use the law of cosines to solve this problem. The lengths a, b, and c of a triangle can be used in the formula below to solve for angle C, which is the angle opposite the side with length c.

$$c^2 = a^2 + b^2 - 2ab\cos C$$

In this case, *a* and *b* are 11 and 7 units, and *c* is 6 units. The angle α opposite side *c* corresponds to angle *C* in the formula.

$$6^2 = 11^2 + 7^2 - 2(11)(7)\cos\alpha$$
$$36 = 121 + 49 - 154\cos\alpha = 170 - 154\cos\alpha$$
$$154\cos\alpha = 134$$
$$\cos\alpha = \frac{134}{154} \approx 0.870$$
$$\alpha = \arccos 0.870 \approx 29.5°$$

Thus, angle α is approximately 29.5°.

18. For an acute angle x, sin x = 0.6. What is cot x?
    (Rigorous)

    A. $\dfrac{5}{3}$

    B. 0.75

    C. 1.33

    D. 1

Answer: B. 0.75
Using the Pythagorean Identity, it is apparent that $\sin^2 x + \cos^2 x = 1$. Thus, $\cos x = \sqrt{1 - \dfrac{9}{25}} = \dfrac{4}{5}$ and $\cot x = \dfrac{\cos x}{\sin x} = \dfrac{4}{3}$.

19. Determine the rectangular coordinates of the point with polar coordinates (5, 60°).
    (Average)

    A. (0.5, 0.87)

    B. (−0.5, 0.87)

    C. (2.5, 4.33)

    D. (25, 150°)

Answer: C. (2.5, 4.33)
Given the polar point $(r, \theta) = (5, 60)$, the rectangular coordinates can be found as follows: $(x,y) = (r\cos\theta, r\sin\theta) = (5\cos 60, 5\sin 60) = (2.5, 4.33)$.

20. A population P of bacteria doubles in number every hour. Which of the following functions of t in hours best represents the number of bacteria in the population?
    (Rigorous) (Skill 5.2)

    A. P^t

    B. $P(2^t)$

    C. Pt^2

    D. Pe^t

Answer: B. $P(2^t)$

One approach to this problem is to consider some simple numbers. Assume, for instance, that the initial population (time $t = 0$ hours) includes two bacteria. The population at time t is then the following:

t (hours)	P
1	4
2	8
3	16
4	32
5	64
6	128

At this point, you can either attempt to derive an appropriate function or you can test each function listed in the possible choices for the problem. Only answer B yields the numbers in the table; thus, $P(2^t)$ is the correct answer.

21. What is the domain of the function $g(x) = \tan x$?
(Average)

A. $\{x \in \mathbb{R}\}$

B. $\{x \in \mathbb{R} : x \neq \pm n\pi\}$ (*n* odd)

C. $\left\{x \in \mathbb{R} : x \neq \pm \dfrac{n\pi}{2}\right\}$ (*n* odd)

D. $\{\varnothing\}$

Answer: C. $\left\{x \in \mathbb{R} : x \neq \pm \dfrac{n\pi}{2}\right\}$ (*n* odd)

Trigonometric functions are periodic, so we know that any values not in the domain of the tangent function will occur with some periodicity. One approach to the problem is to take a look at the graph of the function to get an idea of its behavior.

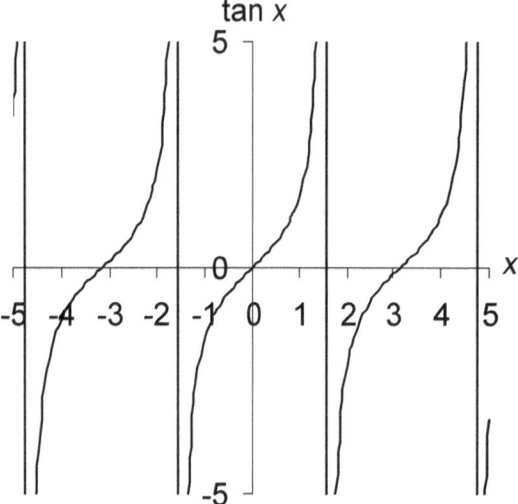

Note that the asymptotes (shown as vertical lines) show up at $-\dfrac{3\pi}{2}, -\dfrac{\pi}{2}, \dfrac{\pi}{2}$, and so on. Again, because the tangent function is periodic, we can say that the vertical asymptotes occur at $x = \pm \dfrac{n\pi}{2}$ for all odd integers *n*. Thus, the domain of the function is all real *x* where $x \neq \pm \dfrac{n\pi}{2}$. We can denote this as shown in answer C above.

22. Find the zeroes of $f(x) = x^3 + x^2 - 14x - 24$
    (Rigorous) (Skill 5.4)

    A. 4, 3, 2

    B. 3, –8

    C. 7, –2, –1

    D. 4, –3, –2

Answer: D. 4, –3, –2
Possible rational roots of the equation $0 = x^3 + x^2 - 14x - 24$ are all the positive and negative factors of 24. By substituting into the equation, we find that –2 is a root, and therefore that $x + 2$ is a factor. By performing the long division $(x^3 + x^2 - 14x - 24)/(x + 2)$, we can find that another factor of the original equation is $x^2 - x - 12$ or $(x - 4)(x + 3)$. Therefore, the zeros of the original function are –2, –3 and 4.

23. Which equation corresponds to the logarithmic statement:
    $\log_x k = m$?
    (Rigorous) (Skill 5.4)

    A. $x^m = k$

    B. $k^m = x$

    C. $x^k = m$

    D. $m^x = k$

Answer: A. $x^m = k$
By definition of log form and exponential form, $\log_x k = m$ corresponds to $x^m = k$.

24. Which expression is equal to $x^4 + 2x^3 - 16x^2 - 2x + 15$ divided by $x + 5$?
    (Rigorous)

    A. $4x^3 + 6x^2 - 32x - 2$

    B. $x^3 - 15x^2 - 5x + 15$

    C. $x^3 - 3x^2 - x + 3$

    D. $5x^3 + 10x^2 - 80x + 75$

Answer: C. $x^3 - 3x^2 - x + 3$

The most direct approach to solving this problem is the use of synthetic division. We can set up the synthetic division as shown below. Inside the division bar, we place the coefficients of the polynomial that we are dividing, and outside the division bar, we write the solution ($x = -5$) corresponding to the factor $x + 5$.

```
-5 | 1   2   -16   -2   15
   |
   |_____
```

Next, we start by carrying the first coefficient down to the bottom, then we begin performing the algorithm for synthetic division. The first step is shown below, followed by the final result.

```
-5 | 1   2   -16   -2   15
   |
   |_____
     1
```

```
-5 | 1   2   -16   -2   15
   |      -5   15    5  -15
   |_____
     1   -3   -1    3    0
```

We can now write the resulting polynomial, which is the solution to the problem: $x^3 - 3x^2 - x + 3$. Of course, an alternative (but tedious) approach to solving this problem is to multiply $x + 5$ by each potential answer to see which product is equal to the original polynomial given in the problem.

25. Solve for *x*: $10^{x-3} + 5 = 105$.
  (Rigorous) (Skill 5.4)

    A.   3

    B.   10

    C.   2

    D.   5

Answer: D. 5

Simplify: $10^{x-3} = 100$. Taking the logarithm to base 10 of both sides yields $(x-3)\log_{10} 10 = \log_{10} 100$. Thus, $x - 3 = 2$ and $x = 5$.

26. Find the inverse of the function $f(x) = 2x^2 - 3$.
    (Average)

    A. $f^{-1}(x) = \sqrt{\dfrac{x+3}{2}}$

    B. $f^{-1}(x) = 2x^2 + 3$

    C. $f^{-1}(x) = \sqrt{2x^2 + 3}$

    D. The function does not have an inverse

Answer: D. The function does not have an inverse

A function has an inverse if and only if it is one-to-one. A one-to-one function is a function that satisfies both the horizontal and vertical line tests (if it is a function, then by definition it already satisfies the vertical line test). Let's take a look at the graph of this function.

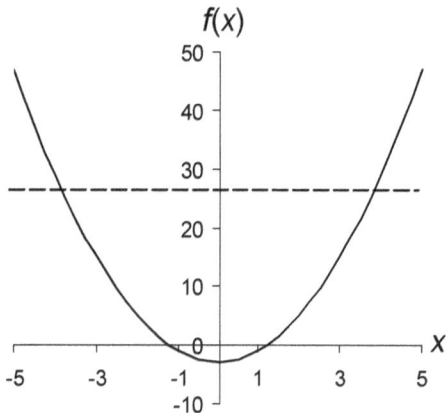

Note that the function does not satisfy the horizontal line test (two unique x values produce the same value of the function f). Thus, this function has no inverse, and answer D is correct.

27. Which of the following represents $f \circ g$, where $f(x) = 3x^2 + 1$ and $g(x) = 2 \sin x - 1$?
    (Average) (Skill 5.5)

    A. $2\sin(3x^2+1)-1$

    B. $6\sin x - 2$

    C. $3\sin x + 1$

    D. $12\sin^2 x - 12\sin x + 4$

Answer: D. $12\sin^2 x - 12\sin x + 4$

The expression $f \circ g$ represents the composition of functions $f(g(x))$. Thus, we can find $f \circ g$ by simply substituting $g(x)$ for the argument in $f(x)$.

$$f \circ g = f(g(x)) = 3(2\sin x - 1)^2 + 1 = 3(4\sin^2 x - 4\sin x + 1) + 1$$
$$f \circ g = f(g(x)) = 12\sin^2 x - 12\sin x + 3 + 1 = 12\sin^2 x - 12\sin x + 4$$

Thus, the composite in this case is $12\sin^2 x - 12\sin x + 4$.

28. Which of the following functions does not have an inverse?
    (Average)

    A. x^3

    B. $\ln \dfrac{x}{2}$

    C. e^{x^2}

    D. $\dfrac{1}{x}$

Answer: C. e^{x^2}

For a function to have an inverse, it must be one-to-one. A simple way to test a function is to apply the horizontal and vertical line tests. If there exists either a horizontal or vertical line that intersects the plot of the function (or relation, more generally), then it is not one-to-one and therefore does not have an inverse. For answers A, B, and D, the functions all pass the horizontal and vertical line tests. Note below, however, that the function in answer C fails the horizontal line test.

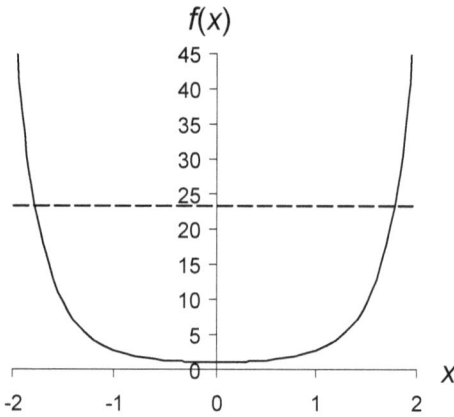

29. **Find the following limit:** $\lim\limits_{x \to 2} \dfrac{x^2 - 4}{x - 2}$

    (Average)

    A. 0

    B. Infinity

    C. 2

    D. 4

Answer: D. 4

First, factor the numerator and cancel the common factor to get the limit.

$$\lim_{x \to 2} \frac{x^2 - 4}{x - 2} = \lim_{x \to 2} \frac{(x - 2)(x + 2)}{(x - 2)} = \lim_{x \to 2} (x + 2) = 4$$

30. **Find the following limit:** $\lim\limits_{x \to 0} \dfrac{\sin 2x}{5x}$

    (Rigorous)

    A. Infinity

    B. 0

    C. 1.4

    D. 1

Answer: C. 1.4

Since substituting $x = 0$ will give an undefined answer, we can use L'Hospital's rule and take derivatives of both the numerator and denominator to find the limit.

$$\lim_{x \to 0} \frac{\sin 2x}{5x} = \lim_{x \to 0} \frac{2 \cos 2x}{5} = \frac{2}{5} = 1.4$$

31. The radius of a spherical balloon is increasing at a rate of 2 feet per minute. What is the rate of increase of the volume when the radius is 4 feet?
    (Rigorous)

    A. 4 feet³/minute

    B. 32π feet³/minute

    C. 85.3π feet³/minute

    D. 128π feet³/minute

Answer: D. 128π feet³/minute
First, note that the volume V of a sphere in terms of its radius r is the following.

$$V = \frac{4}{3}\pi r^3$$

We want to calculate the rate of change of the volume with respect to time $\left(\frac{dV}{dt}\right)$ in terms of the rate of change of the radius with respect to time $\left(\frac{dr}{dt}\right)$. First, use the chain rule to rewrite the expression for the rate of change of the volume.

$$\frac{dV}{dt} = \frac{dV}{dr}\frac{dr}{dt}$$

Now, calculate the derivative of the volume with respect to the radius.

$$\frac{dV}{dr} = 4\pi r^2$$

Then,

$$\frac{dV}{dt} = 4\pi r^2 \frac{dr}{dt} = 4\pi r^2 \left(2\frac{\text{feet}}{\text{minute}}\right) = 8\pi r^2 \frac{\text{feet}}{\text{minute}}$$

When the radius is 4 feet, the rate of change of the volume is as follows.

$$\frac{dV}{dt} = 8\pi(4\text{ feet})^2 \frac{\text{feet}}{\text{minute}} = 128\pi \frac{\text{feet}^3}{\text{minute}}$$

Thus, the rate of change of the volume when the radius is 4 feet is 128π cubic feet per minute.

32. What is the maximum value of the function $f(x) = -3x^2 - 5$?
    (Average)

    A. −5

    B. −3

    C. 1

    D. 5

Answer: A. −5

First, consider the graph of this function, which is shown below.

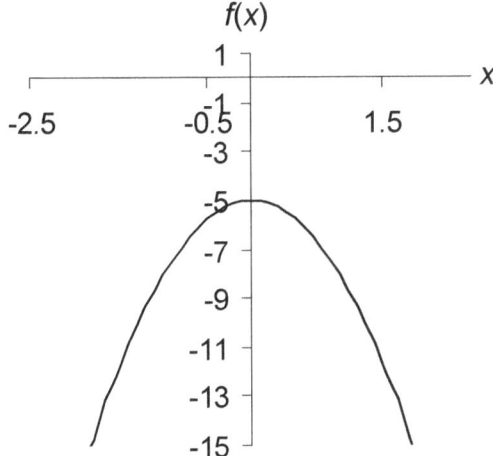

Obviously, the function has a single maximum value. To find the maximum value, first determine the location of the critical point. The critical points correspond to locations where the slope of the function is zero; to find these points, set the first derivative of the function equal to zero, and solve for the variable.

$$f'(x) = \frac{d}{dx}(-3x^2 - 5) = -6x$$
$$-6x = 0$$
$$x = 0$$

Now, to find the maximum value, substitute $x = 0$ into the function $f(x)$.

$$f(0) = -3(0)^2 - 5 = -5$$

33. Find the antiderivative for the function $y = e^{3x}$.
    (Rigorous)

    A. $3x(e^{3x}) + C$

    B. $3(e^{3x}) + C$

    C. $1/3(e^x) + C$

    D. $1/3(e^{3x}) + C$

Answer: D. $1/3(e^{3x}) + C$

Use the rule for integration of functions of e ($e^x dx = e^x + C$) along with definition of a new variable $u = 3x$. The result is answer D.

34. Evaluate $\int_0^2 (x^2 + x - 1)dx$
    (Rigorous)

    A.   11/3

    B.   8/3

    C.   −8/3

    D.   −11/3

Answer: B. 8/3

Use the fundamental theorem of calculus to find the definite integral: given a continuous function f on an interval $[a,b]$, then $\int_a^b f(x)dx = F(b) - F(a)$, where F is an antiderivative of f.

$$\int_0^2 (x^2 + x - 1)dx = (\frac{x^3}{3} + \frac{x^2}{2} - x)$$

Evaluate the expression at $x = 2$, at $x = 0$, and then subtract to get 8/3 + 4/2 − 2 − 0 = 8/3.

35. What conclusion can be drawn from the graph below?

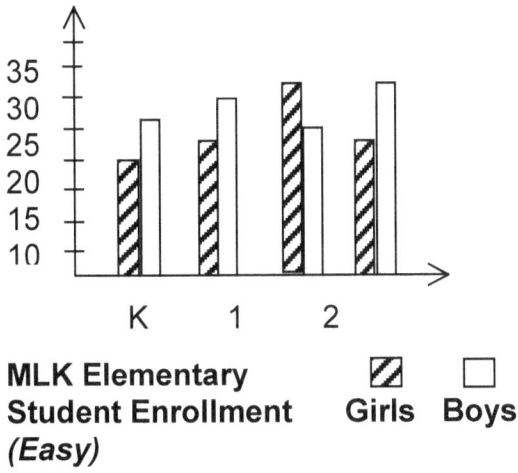

MLK Elementary Student Enrollment   ▨ Girls   ☐ Boys
(Easy)

- A. The number of students in first grade exceeds the number in second grade
- B. There are more boys than girls in the entire school
- C. There are more girls than boys in the first grade
- D. Third grade has the largest number of students

Answer: B. There are more boys than girls in the entire school
In Kindergarten, first grade and third grade, there are more boys than girls. The number of extra girls in grade two is more than compensated by the extra boys in all the other grades put together.

36. The pie chart below shows sales at an automobile dealership for the first four months of a year. What percentage of the vehicles were sold in April?
(Easy)

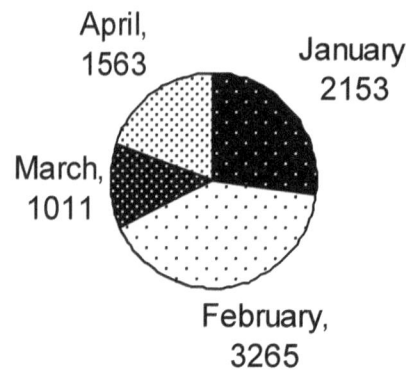

- A. More than 50%
- B. Less than 25%
- C. Between 25% and 50%
- D. None

Answer: B. Less than 25%
It is clear from the chart that the April segment covers less than a quarter of the pie.

37. What is the standard deviation of the following sample?

{1.46, 1.55, 1.55, 1.57, 1.89, 2.01, 2.09}

(Average)

- A.   0.010
- B.   0.066
- C.   0.26
- D.   1.73

Answer: C. 0.26
Notice first that the problem stated that the data set is a *sample*. Thus, we need to use sample statistics (as will be discussed further). To start, we must calculate the mean $\bar{x}$ of the data.

$$\bar{x} = \frac{1.46+1.55+1.55+1.57+1.89+2.01+2.09}{7} \approx 1.73$$

The standard deviation is the square root of the sample variance, s^2, which is given below.

$$s^2 = \frac{1}{n-1}\sum_i (x_i - \bar{x})^2$$

$$s^2 = \frac{1}{6}\left[(1.46-1.73)^2 + (1.55-1.73)^2 + \cdots + (2.09-1.73)^2\right]$$

$$s^2 \approx \frac{1}{6}[0.073+0.032+\cdots+0.130] \approx 0.066$$

$$s = \sqrt{s^2} \approx \sqrt{0.066} \approx 0.26$$

The standard deviation *s* is then 0.26. The correct answer is thus C.

38. Which measure of central tendency best characterizes the data set below?

Value	Frequency
1	2
2	5
3	7
4	3
5	2
6	1
7	1

(Rigorous)

A. Mean

B. Median

C. Both the mean and median are the same

D. None of the above

Answer: A. Mean

By plotting the data on a bar graph or histogram, we can visualize the shape of the data distribution.

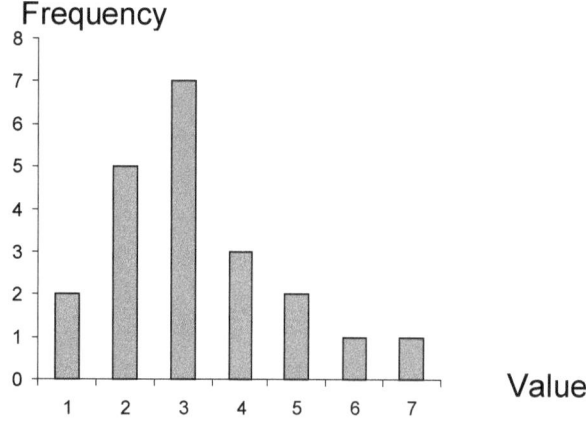

Because the distribution is skewed rather than symmetric, it is not immediately clear that the mean or median are the best choices in this situation. The median is 4 (the data set contains a total of 21 values). We can calculate the mean using a weighted average as shown below.

$$\bar{x} = \frac{1\cdot 2 + 2\cdot 5 + 3\cdot 7 + \cdots + 7\cdot 1}{2+5+7+\cdots+1} = \frac{68}{21} \approx 3.24$$

The mean in this case is a little closer to the peak. Given the shape of the distribution, the mean is therefore a better choice than the median. Choice A is therefore correct.

39. **The height of people in a certain city is a normally distributed random variable. If a person is chosen from the city at random, what is the probability that he or she has a height greater than the mean of the distribution?**
    (Average)

    A.   0.25

    B.   0.5

    C.   0.75

    D.   Not enough information

Answer: B. 0.5
A normal distribution is symmetric about the mean, as shown below.

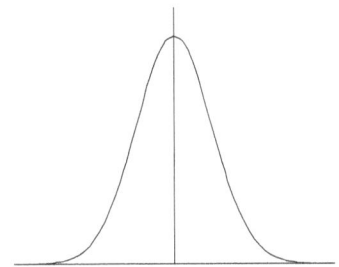

As such, the probability that a particular value is greater than (or less than) the mean is 0.5, since half the probability density is on one side of the mean, and the other half is on the other side.

40. If there are three people in a room, what is the probability that at least two of them will share a birthday? (Assume a year has 365 days.)
    (Rigorous)

    A.   0.67

    B.   0.05

    C.   0.008

    D.   0.33

Answer: C. 0.008
The best way to approach this problem is to use the fact that the probability of an event plus the probability of the event not happening is unity. First, find the probability that no two people will share a birthday and then subtract that value from one. The probability that two of the people will not share a birthday is 364/365 (since the second person's birthday can be one of the 364 days other than the birthday of the first person). The probability that the third person will also not share either of the first two birthdays is (364/365)(363/365) = 0.992. Therefore, the probability that at least two people will share a birthday is 1 – 0.992 = 0.008.

41. **How many different five-card hands containing three aces and two kings can be drawn from a standard 52-card deck?**
    (Rigorous)

    A. 6

    B. 16

    C. 24

    D. 2,598,960

Answer: C. 24
A standard 52-card deck contains four kings and four aces. The number of five-card hands containing three aces and two kings can be found as the product of the number of combinations of three aces and the number of combinations of two kings. Thus, the solution N is the following, where $\binom{n}{k}$ is the number of combinations of n objects taken k at a time.

$$N = \binom{4}{3} \cdot \binom{4}{2} = 4 \cdot 6 = 24$$

Thus, 24 possible five-card hands have three aces and two kings.

42. **What is the probability that a roll of a six-sided die yields an outcome that is even or greater than three?**
    (Average)

    A. 3/6

    B. 4/6

    C. 5/6

    D. 1

Answer: B. 4/6
The sample space for a single roll of a six-sided die is {1, 2, 3, 4, 5, 6}. The outcomes that satisfy the condition of being even or greater than three are {2, 4, 5, 6}. Thus, four out of a total of six possible outcomes result in a "successful" trial. The corresponding probability is then 4/6 = 2/3, or approximately 0.67.

43. Find the sum of the following matrices is

$$\begin{pmatrix} 6 & 3 \\ 9 & 15 \end{pmatrix} \begin{pmatrix} 4 & 7 \\ 1 & 0 \end{pmatrix}$$

(Easy)

A. $\begin{pmatrix} 10 & 10 \\ 10 & 15 \end{pmatrix}$

B. $\begin{pmatrix} 13 & 7 \\ 9 & 16 \end{pmatrix}$

C. 45

D. $\begin{pmatrix} 20 \\ 25 \end{pmatrix}$

Answer: A. $\begin{pmatrix} 10 & 10 \\ 10 & 15 \end{pmatrix}$

Two matrices with the same dimensions are added by adding the corresponding elements. In this case, element 1,1 (i.e. row 1, column 1) of the first matrix is added to element 1,1 of the second matrix; element 2,1 of the first matrix is added to element 2,1 of the second matrix; and so on for all four elements.

44. The product of two matrices can be found only if
(Easy)

    A.   The number of rows in the first matrix is equal to the number of rows in the second matrix

    B.   The number of columns in the first matrix is equal to the number of columns in the second matrix

    C.   The number of columns in the first matrix is equal to the number of rows in the second matrix

    D.   The number of rows in the first matrix is equal to the number of columns in the second matrix

Answer: C. The number of columns in the first matrix is equal to the number of rows in the second matrix
The number of columns in the first matrix must equal the number of rows in the second matrix because the process of multiplication involves multiplying the elements of every row of the first matrix with corresponding elements of every column of the second matrix.

45. Which of the following properties does not apply to matrix multiplication?
(Easy)

    A.   Associativity

    B.   Commutativity

    C.   Distributivity

    D.   All of the above

Answer: B. Commutativity
Matrix multiplication obeys associativity and distributivity but not commutativity, as shown below.

$$\begin{pmatrix} a & b \\ c & d \end{pmatrix} \begin{pmatrix} e & f \\ g & h \end{pmatrix} = \begin{pmatrix} ae-bg & af-bh \\ ce-dg & cf-dh \end{pmatrix}$$

$$\begin{pmatrix} e & f \\ g & h \end{pmatrix} \begin{pmatrix} a & b \\ c & d \end{pmatrix} = \begin{pmatrix} ae-cf & be-df \\ ag-ch & bg-dh \end{pmatrix}$$

Obviously, the two products are different.

46. Which of the following best describes the translation matrix below for arbitrary points (x_1, y_1)?

$$\begin{pmatrix} 1 & 0 \\ 0 & -1 \end{pmatrix} \begin{pmatrix} x_1 \\ y_1 \end{pmatrix} = \begin{pmatrix} x_2 \\ y_2 \end{pmatrix}$$

(Average)

    A.   Rotation

    B.   Translation

    C.   Reflection

    D.   Dilation

Answer: C. Reflection
Perform the matrix multiplication to see what happens to the point (x_1, y_1).

$$\begin{pmatrix} 1 & 0 \\ 0 & -1 \end{pmatrix} \begin{pmatrix} x_1 \\ y_1 \end{pmatrix} = \begin{pmatrix} x_1 \\ -y_1 \end{pmatrix}$$

Notice, then, that the result is a corresponding point on the opposite side of the *x*-axis. Thus, this translation matrix reflects points across the *x*-axis.

47. The Fibonacci numbers are defined recursively using the expression $F_{i+1} = F_i + F_{i-1}$. If $F_1 = 0$ and $F_2 = 1$, what is F_8?
(Easy)

    A.   0

    B.   1

    C.   13

    D.   21

Answer: C. 13
The recursive expression simply states that each Fibonacci number is simply the sum of the previous two numbers. Thus, if we start with zero and one, then the Fibonacci numbers are {0, 1, 1, 2, 3, 5, 8, 13, 21,…}. On the basis of these results, F_8 is 13.

48. On which of the following sets is multiplication *not* symmetric?
(Average)

   A.   Real numbers

   B.   Complex numbers

   C.   Polynomials

   D.   Matrices

Answer: D. Matrices
A binary relation R is symmetric on a set if, for all a and b in a set, both aRb and bRa have the same truth value. For real numbers, complex numbers and polynomials with members a and b in each case, ab is always equal to ba. For matrices, however, ab is not always equal to ba for given matrices a and b. Thus, the correct answer is D.

49. Which of the following can be used to prove that $0.\overline{9} = 1$?
(Rigorous)

    A.   Arithmetic series

    B.   Geometric series

    C.   Fibonacci sequence

    D.   Irrational numbers

Answer: B. Geometric series

The number 0.9999… can be represented using a geometric series, as shown below.

$$0.\overline{9} = 9 \cdot \frac{1}{10} + 9 \cdot \frac{1}{100} + 9 \cdot \frac{1}{1,000} + \cdots = 9\left[\left(\frac{1}{10}\right)^1 + \left(\frac{1}{10}\right)^2 + \left(\frac{1}{10}\right)^3 + \cdots\right]$$

$$0.\overline{9} = \frac{9}{10}\left[\left(\frac{1}{10}\right)^0 + \left(\frac{1}{10}\right)^1 + \left(\frac{1}{10}\right)^2 + \cdots\right]$$

But the series in brackets is simply a geometric series. We can use the following, which applies to $r < 1$, to write the above series in closed form.

$$1 + r + r^2 + r^3 + \cdots = \frac{1}{1-r}$$

Thus,

$$0.\overline{9} = \frac{9}{10}\left(\frac{1}{1-0.1}\right) = \frac{9}{10}\left(\frac{1}{0.9}\right) = \frac{9}{10}\left(\frac{10}{9}\right) = 1$$

A geometric series can therefore be used to prove that $0.\overline{9} = 1$.

50. What is the shortest path between points A and B?

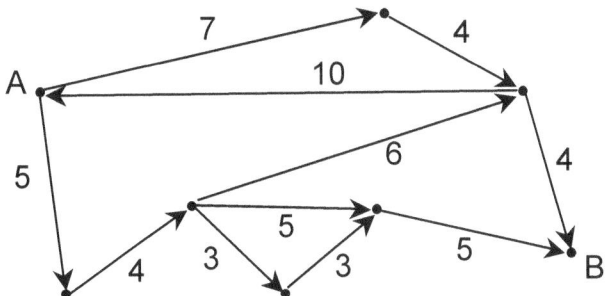

(Easy)

A. 11

B. 14

C. 15

D. 19

Answer: C. 15
When finding the shortest path, be sure to follow the paths in the appropriate directions. Only several direct (that is, non-circular) paths are possible; of these, the shortest path has a length of 15. This path is shown below.

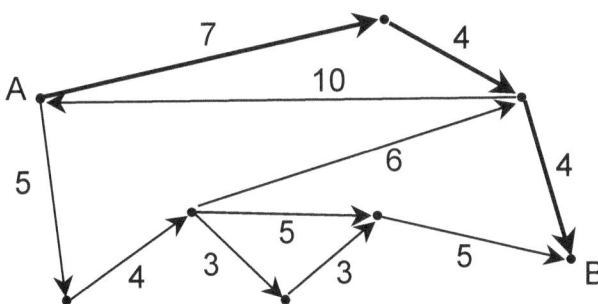

Answer Key

1.	D		26.	D
2.	D		27.	D
3.	C		28.	C
4.	A		29.	D
5.	A		30.	C
6.	C		31.	D
7.	A		32.	A
8.	B		33.	D
9.	B		34.	B
10.	C		35.	B
11.	A		36.	B
12.	B		37.	C
13.	B		38.	A
14.	B		39.	B
15.	A		40.	C
16.	C		41.	C
17.	C		42.	B
18.	B		43.	A
19.	C		44.	C
20.	B		45.	B
21.	C		46.	C
22.	D		47.	C
23.	A		48.	D
24.	C		49.	B
25.	D		50.	C

Rigor Table

	Easy 20%	Average 40%	Rigorous 40%
Questions	7, 14, 16, 35, 36, 43, 44, 45, 47, 50	1, 2, 3, 4, 8, 9, 10, 11, 19, 21, 26, 27, 28, 29, 32, 37, 39, 42, 46, 48	5, 6, 12, 13, 15, 17, 18, 20, 22, 23, 24, 25, 30, 31, 33, 34, 38, 40, 41, 49

www.ingramcontent.com/pod-product-compliance
Lightning Source LLC
LaVergne TN
LVHW061317060426
835507LV00019B/2196